ROCKS AND MINERALS

Molly Blaisdell

Perfection Learning®

TABLE OF CONTENTS

1

WHAT ARE ROCKS AND MINERALS?

Rocks are *everywhere*. The whole Earth is made of rocks. The inside of the Earth is melted rock. The outside of the Earth is the crust. It is made of rocks too. Then there are the mountains, the riverbeds, and the sand at the beach. They are rocks too.

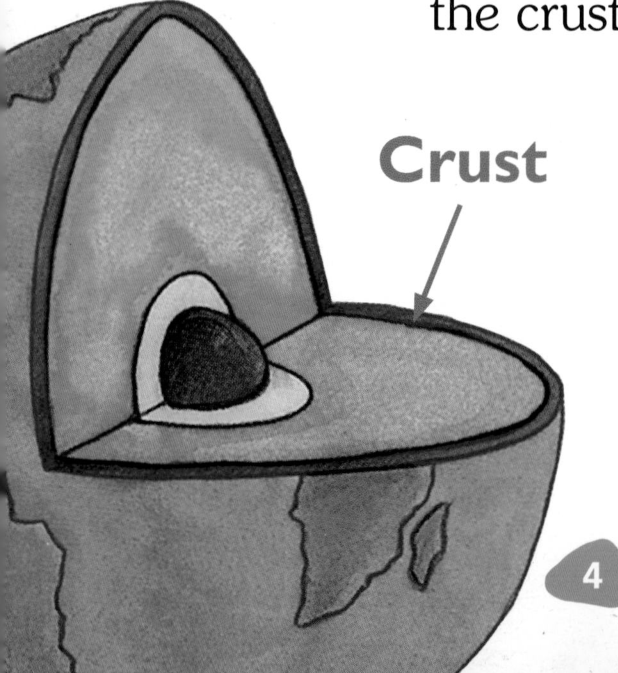

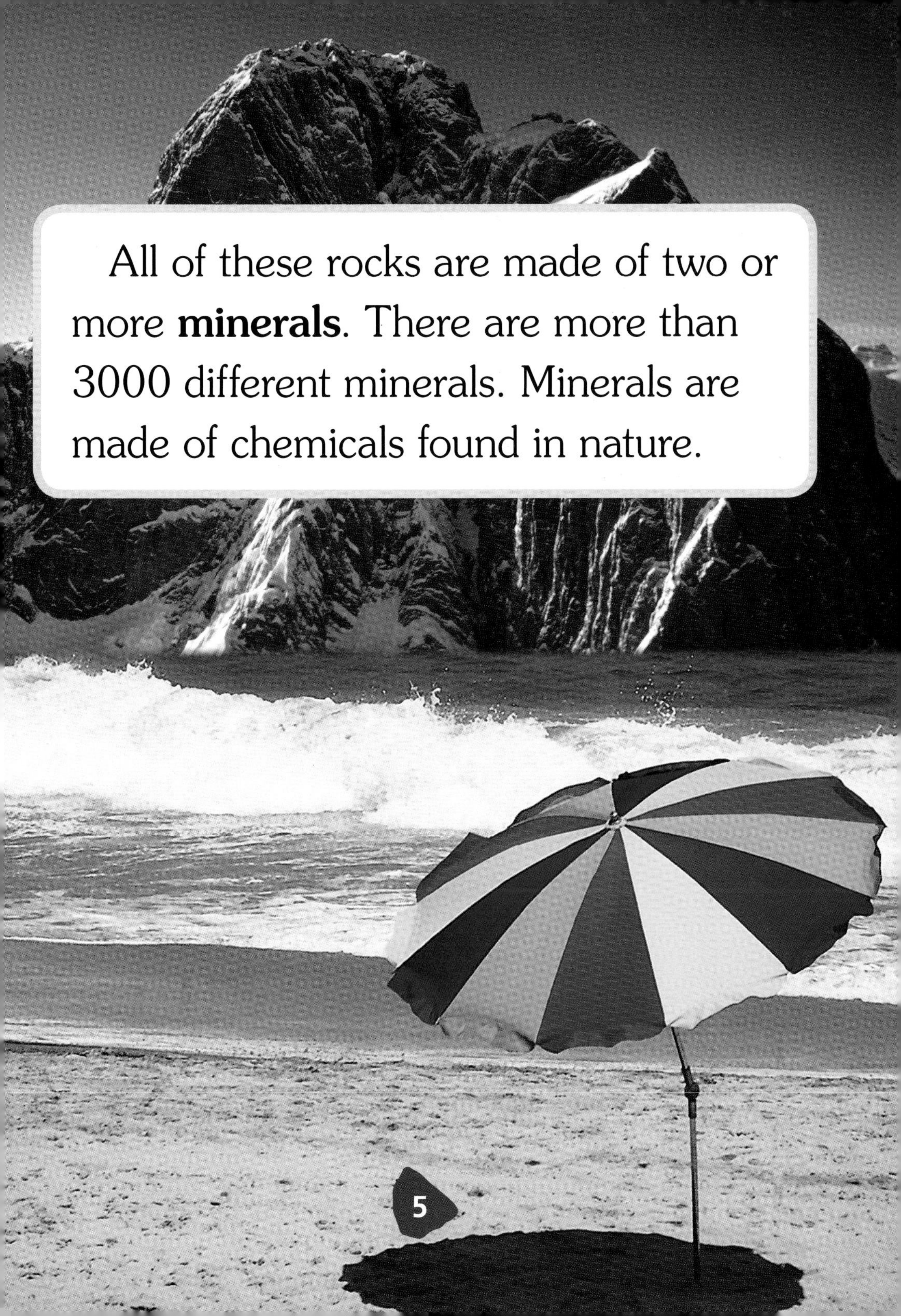

All of these rocks are made of two or more **minerals**. There are more than 3000 different minerals. Minerals are made of chemicals found in nature.

Uses of Rocks and Minerals

Rocks and minerals have been used throughout history. Rocks were the first tools. Pyramids were built with boulders. Walls and castles were built with rocks.

Rock Words

mountain	huge	part of the Earth's surface
boulder	large	taller than a person
stone	medium	fits in two hands
pebble	small	size of a grape
sand	tiny	like a grain of rice
dust	minute	fine powder

Minerals have also been used to make many things. People prize minerals called ***gems*** for their beauty. Gems are used to make jewelry. Gold, silver, and copper are minerals used to make coins. **Silicon** crystals are sliced to make mineral chips. They are used in computers.

A Rare Mineral

The largest cut diamond in the world is called the Star of Africa. It weighs about 1.25 pounds. It is in the Tower of London. This gem is part of a king's **scepter**.

2

KINDS OF ROCKS

There are three kinds of rocks—**igneous**, **sedimentary**, and **metamorphic**. New rocks are always forming on the Earth. This is the **rock cycle**.

Igneous Rock

Beneath the Earth's rocky crust, it is very hot. Rocks melt. The melted rock is called ***magma***. When the magma cools inside the Earth's crust, new rocks form. Sometimes magma erupts from volcanoes. This melted rock is called ***lava***. The lava hardens, and new rocks form. These rocks are called *igneous rocks*. *Igneous* means "fire."

Sedimentary Rock

Over many years, rocks are broken down into dirt and sand by wind and rain. These broken-down rocks are called ***sediments***. Layers of sediment settle in riverbeds or oceans. The layers are heavy. They press together and form sedimentary rock.

Metamorphic Rock

Time passes. More layers form. Many layers of sediment are very heavy and press on the rocks below. Heat and **pressure** cause sedimentary rocks to be changed into metamorphic rock.

Layers of sedimentary rock keep forming above. Metamorphic rock is pushed deeper. It is melted again by the heat at the Earth's center and becomes magma. When magma cools, it becomes igneous rock again. The rock cycle starts over.

Igneous
wind and rain
The
Rock Cycle
melts and cools
Metamorphic
Sedimentary
heat and pressure

3

KINDS OF MINERALS

There are three kinds of minerals—**metals**, **semimetals**, and **nonmetals**.

Metals are bendable, soft minerals. They **conduct electricity**. Copper is a metal.

Semimetals are small, rounded lumps of mineral. They do not conduct much electricity.

Nonmetals do not conduct any electricity. These minerals tend to form crystals.

Metal

There are more than 3000 minerals. Scientists ask six questions to help them **identify** minerals.

1. **What is the shape of the mineral?** Some minerals tend to have a plant shape. Others are found as cubes. The special shape of a mineral makes it easy to identify.

2. **What does the surface look like?** Many minerals reflect light. A mineral can look sparkly, greasy, bright, pearly, or brilliant. The surface of a mineral makes it easy to identify.

3. What color is the mineral?
Many minerals have a unique color. This makes them easy to identify.

4. Can you draw with the mineral?
Some minerals will make a mark on paper just like a crayon. The mark of a mineral makes it easy to identify.

5. How hard is the mineral?

Talc is the softest mineral. Diamond is the hardest. Hardness is tested by scratching the mineral. The hardness of a mineral makes it easy to identify.

6. How does a mineral break?

Most minerals break in a special way. The way a mineral breaks makes it easy to identify.

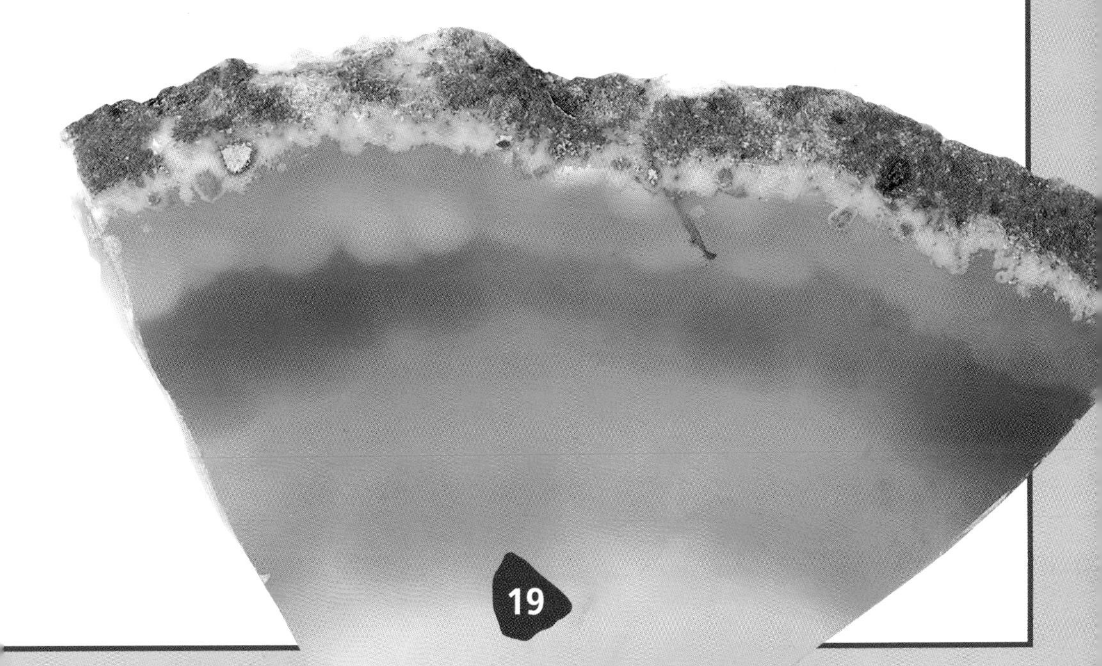

Rock Collectors

Many people enjoy collecting rocks and minerals. You may enjoy creating a rock collection. Place small rocks and minerals in an egg carton. Use rock and mineral guides to help you identify your samples.

Glossary

conduct	to carry along
electricity	form of energy caused by electrons moving through a mineral
gem	mineral that has been cut and polished and is prized for its beauty
identify	to show what something is
igneous	rocks formed with great heat, such as from a volcano
lava	melted rock erupting from a volcano
magma	melted rock inside the Earth

metal	group of similar minerals that conduct heat and electricity
metamorphic	relating to changes caused by great heat or pressure
mineral	substances on Earth that are not plants or animals
nonmetals	group of similar minerals that do not conduct heat and electricity
pressure	steady force
rock	solid that is a mix of minerals
rock cycle	ongoing formation of new rocks

scepter rod that is a symbol of royal power

sediment solid material such as dirt and sand that settle at the bottom of water

sedimentary formed by layers of sediments

semimetals group of similar minerals that form as round lumps and conduct little electricity

silicon hard, dark gray mineral

Index